# THE WEATHER AND US

## WEATHER REPORT

Ted O'Hare

Rourke
Publishing LLC
Vero Beach, Florida 32964

© 2003 Rourke Publishing LLC

All rights reserved. No part of this book may be reproduced or utilized in any form or by any means, electronic or mechanical including photocopying, recording, or by any information storage and retrieval system without permission in writing from the publisher.

www.rourkepublishing.com

PHOTO CREDITS: Page 7 courtesy of South Dakota Tourism; Page 15, NASA; all other photos © Lynn M. Stone

Title page: *Snow-covered hills make wintertime fun.*

Editor: Henry Rasof

Cover and page design by Nicola Stratford

## Library of Congress Cataloging-in-Publication Data

O'Hare, Ted, 1961-
 The weather and us / Ted O'Hare
 v. cm.— (Weather report)
Contents: Weather around us — Changing weather — What causes weather? — Climate — The atmosphere — Wind — Hot and cold places — Wet and dry places. Includes bibliographical references and index.
ISBN 1-58952-574-4 (hardcover)
1. Weather—Juvenile literature. 2. Climatology—Juvenile literature. (1. Weather. 2. Climatology.
QC981.3 .O43 2002          2002151638
551.5—dc21

Printed in the USA

# Table of Contents

| | |
|---|---|
| Weather around Us | 5 |
| Changing Weather | 6 |
| What Causes Weather? | 9 |
| Climate | 10 |
| The Atmosphere | 14 |
| Wind | 16 |
| Hot and Cold Places | 19 |
| Wet and Dry Places | 20 |
| The Weather and Us | 22 |
| Glossary | 23 |
| Index | 24 |
| Further Reading/Websites to Visit | 24 |

# Weather around us

Weather is important to everyone. Weather is what it is like outdoors on any day. Weather may determine where we live and how we live. It affects the way we dress and what we can do outside.

Weather also has much to do with how we feel. We talk a lot about the weather, but we can't do much about it. Weather changes itself.

*Good weather can make a field trip great fun!*

# Changing weather

Weather changes because of the way **air masses** meet. Air masses are great "pieces" of air. They may be warm or cold, or they may be wet or dry.

When two air masses meet, there will be a change in weather conditions. When two air masses meet, a **front** is formed. The front often brings rain or snow.

*This wet, fast-moving air mass promises a change in weather conditions.*

# What causes weather?

Weather comes about because of different things. One is the amount of sunshine that a place gets. Another is the height, or **altitude**, of a place. Generally, the higher a place's altitude, the cooler the air.

Mountains near a place affect the weather. So does the closeness of a place to large bodies of water. The direction and force of wind also affect weather.

*Mountainous areas have a direct impact on weather.*

# Climate

How is **climate** different from weather? Climate is the weather that a place has over a long period of time. A day's weather does not give us a good idea of a place's climate. A desert may have a rainy day from time to time, but overall a desert has a dry climate.

Things that affect weather also affect climate. For example, oceans affect a place's weather. They also may give seaside places a wetter, warmer climate than inland places.

*Low, wet clouds from the Pacific Ocean help keep Oregon's coastal climate mild.*

*Florida's warm climate has made many of its cities among the fastest growing in North America.*

*Winter morning mist adds moisture to Arizona, but the desert remains a basically dry climate.*

# The atmosphere

All weather happens in the earth's **atmosphere**. This is the blanket of air that surrounds the earth. The atmosphere is made up of layers. The air we breathe is the layer of air closest to the earth's surface.

The higher up one goes in the atmosphere, the "thinner" the air becomes. This is because the air contains less **oxygen**. Breathing becomes more difficult. Passengers in an airplane are able to breathe normally because the air in the cabin is **pressurized**.

*The Space Shuttle streaks upward through the earth's atmosphere.*

# Wind

Wind is just moving air. We can feel wind but cannot see it. Wind conditions are part of our weather. When wind blows, the air usually feels cooler than it really is. But sometimes a hot wind can make things seem even hotter.

Wind is caused by differences in the temperatures of air masses. Wind speed several hundred feet above the earth may be stronger than wind speed on the ground. Strong winds can cause damage. Some windstorms, hurricanes, and tornadoes can be very dangerous.

*Winds of 231 miles (372 kilometers) an hour sometimes whirl across Mount Washington in New Hampshire.*

# Hot and cold places

The coldest places on earth are near the North and South poles. Because they are so cold, very few people live in these places. The poles are at the "ends" of the earth. The lowest temperature recorded was in Antarctica. This was minus 128 degrees F (minus 88.9 degrees C).

Some of the world's hottest places are near the **equator**. The equator is an imaginary line around the earth's middle. A scorching temperature of 136 degrees F (57.7 degrees C) has been recorded in North Africa's Sahara Desert.

*The world's coldest continent is Antarctica.*

# Wet and dry places

Rain forests nearest the equator are the wettest places on land. Some of these places have nearly an inch of rain every day. On America's northwest coast there are **temperate** rain forests. These are in Washington, Alaska, and parts of British Columbia.

**Deserts** are dry places. They receive less than 10 inches (25 cm) of rain in a year! The Atacama Desert in Chile has so little rainfall each year that it cannot be measured.

*Warm, drizzly rain forests are the wettest places on the earth's land surface.*

# The weather and us

Most people in the world do not live where the climate is extreme. Would you choose to live near Death Valley, California? In 1913 a temperature of 134 degrees F (56.6 C) was recorded there.

Most people live in climate that is considered temperate. Even within temperate climates there can be vast differences in temperatures and weather.

# Glossary

**air masses** (AER MASSUZ) — great pieces of air in the atmosphere

**altitude** (AL tuh tood) — the height of something above ground

**atmosphere** (AT muss feer) — the "blanket" of air around the earth

**climate** (KLY mut) — the type of weather conditions that any place has over a long period of time

**deserts** (DEZZ urtz) — places where little or no rainfall occurs

**equator** (ee KWAY tur) — the imaginary line around the earth's middle

**front** (FRUHNT) — the boundary between two different air masses

**oxygen** (OKS uh jen) — a colorless gas found in the atmosphere

**pressurized** (PRESS yur eyzd) — normal conditions for air in a place where the air is not normal

**temperate** (TEM pur ut) — neither too hot nor too cold; moderate

# Index

air masses   6, 16
altitude   9
Antarctica   19
Atacama Desert   20
atmosphere   14
climate   10, 22
Death Valley   22
desert   10, 19, 20
equator   19
front   6
mountains   9
oxygen   14
pressurized   14
rain forests   20
Sahara Desert   19
sunshine   9
wind   9, 16

## Further Reading

Bundey, Nikki. *Drought and the Earth*. Minneapolis: Carolrhoda, 2000.
Parker, Janice. *The Science of Weather*. Milwaukee: Gareth Stevens, 2000.
*Weather Watch*. Danbury, CT: Grolier Educational, 2000. Twelve-book series.

## Websites To Visit

www.pbs.org/journeytoamazonia/
www.nws.noaa.gov/om/reachout/kidspage.html
www.usatoday.com/weather/basics/wworks0.htm

## About The Author

Ted O'Hare is an author and editor of children's information books. He divides his time between New York City and a home upstate.